Housing
87

蜜蜂在家里

Bees at Home

Gunter Pauli

[比]冈特·鲍利 著

[哥伦]凯瑟琳娜·巴赫 绘

唐继荣 译

上海远东出版社

丛书编委会

主　任：田成川

副主任：何家振　闫世东　林　玉

委　员：李原原　翟致信　靳增江　史国鹏　梁雅丽
　　　　任泽林　陈　卫　薛　梅　王　岢　郑循如
　　　　彭　勇　王梦雨

特别感谢以下热心人士对童书工作的支持：

匡志强　宋小华　解　东　厉　云　李　婧　庞英元
李　阳　刘　丹　冯家宝　熊彩虹　罗淑怡　旷　婉
杨　荣　刘学振　何圣霖　廖清州　谭燕宁　王　征
李　杰　韦小宏　欧　亮　陈强林　陈　果　寿颖慧
罗　佳　傅　俊　白永喆　戴　虹

目录

Contents

鹦鹉产蛋的时候到了。随着森林被砍伐，树冠几乎全被毁了，她能找到一棵适合筑巢的树实属幸运。但当她回到衬有柔软树叶和羽毛的树洞里面时，鹦鹉发现一只蜂王在同样的位置放松享受。

"蜂王，恕我直言，"鹦鹉说，"是我先到这儿的！"

It is time for a parrot to lay her eggs. As the forest has been cut down and the forest canopy all but destroyed, she is lucky to have found a suitable tree for her nest. Returning to the hole in the tree with soft leaves and feathers to line it, she finds that a queen bee is making herself at home in the very same spot. "With all due respect to you, Queen, " says the parrot, "I was here first!"

鹦鹉产蛋的时候到了

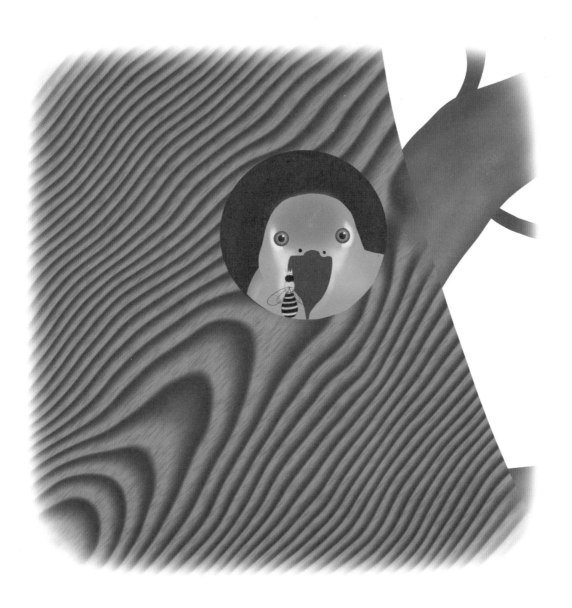

It is time for a parrot to lay her eggs

与蜜蜂分享筑巢地

Sharing a nesting site with bees

"远在你出现之前，我的祖先就住在这片林子里，你怎么可以宣称自己是第一个到这儿的呢？"蜂王回应道。

"好吧！那是很多代之前的事儿啦，没人记得谁真正最先到这儿的。然而，我要告诉你的是，我马上要产蛋了，我无法想象我的孩子与蜜蜂分享筑巢地的情形！"

"How can you claim to have been here first when my ancestors have been living in this forest long before you came along?" responds the bee.

"Well, that was so many generations ago that no one can remember who really was here first. What I can tell you though, is that I need to lay my eggs right now and I cannot imagine my chicks sharing a nesting site with bees!"

"那么，你为什么不能找另一棵有洞的树？我们需要采食这周围大量的花蜜，否则我们的小蜜蜂就不能生存。"蜂王争辩道。

"为了给我们的孩子提供食物，我们需要采食分布在这周围的果实和种子。我相信你会找到另一棵树建你的蜂巢。"鹦鹉说。

"Then why don't you find another tree with a hole in it? We need to feed on the nectar of flowers that grow so abundantly around here or our little ones will not survive," argues the bee.

"And we need to feed on the fruit and seeds that grow around here to provide our chicks with their food. I am sure you will find another tree for your hive," says the parrot.

我相信你会找到另一棵树建你的蜂巢

I am sure you will find another tree for your hive

在这周围几乎没什么树留下来

Hardly any trees left around here

10

"你我都知道，在这周围几乎没什么树留下来了，找到一颗有合适树洞的树几乎是不可能的。"

"这你可不能指责我，"鹦鹉回应道，"你知道，是那些几个世纪前从欧洲来的人把我们这儿的树都砍光了。从那时起，我们就很难找到好的居住地了。"

"You know as well as I do that there are hardly any trees left around here, and that to find one with a suitable hole is just about impossible."

"As long as you don't blame me for that, " replies the parrot. " You know that it was those people from Europe who came here centuries ago who cut down all our trees. We have had a hard time finding a good place to live ever since."

"为什么你不去沙漠，住在仙人掌中？"蜜蜂提议道，"你的羽毛可以保护你免受这些尖刺的伤害，不是吗？"

"为什么你和你成千上万的小工蜂不去呢？你们能轻松地住在仙人掌中，但我尾巴太长就没那么容易。"

"Why don't you go to the desert and live in a cactus?" suggests the bee. "You have feathers to protect you against those sharp spines, don't you?"

"Why don't you and your thousands of little worker bees do that? You lot can easily live in a cactus. It is not so easy for me with my long tail."

你们能住在仙人掌中

you can live in a cactus

让我们尝试共同生活吧

Let's try to live together

"你知道我们在仙人掌中筑巢有多难吗？现在，让我们实际点吧。为什么我们不一起分享这棵稀有的阔叶树上这个完美的树洞呢？让我们尝试共同生活吧。"蜂王提议道。

"你真是不可理喻，蜂王！是我先来的！如果成千上万的蜜蜂在周围嗡嗡作响，从早到晚在蜂巢中进进出出，我们根本没法休息。那我们该在什么时间睡觉呢？"

"Do you have any idea how difficult it is for us to build a hive in a cactus? Now let's be practical here. Why don't we share this perfectly suitable hole in this rare hardwood tree? Let's try to live together," suggests the queen.

"You are really being unreasonable, Queen. I was here first! And with thousands of you buzzing around, entering and leaving the nest all day long, no one will ever get any rest. When are we supposed to sleep?"

"睡觉？没时间睡觉了！作为父母，我们必须确保我们孩子的未来。如果我们能每隔一段时间打个盹，就很幸运了。"

"对啊！自从人们不考虑种植新树就砍倒我们的树木之后，生存变得困难了，我不得不飞得更远去寻找足够的食物。"

"Sleep? There is no time for sleep! As parents we have to secure a future for our little ones. We will be lucky if we manage a short nap once in a while."

"Yes, ever since people cut down our trees without worrying about planting new ones, it has been difficult to survive. I have to roam further and further to get enough food."

我们必须确保我们孩子的未来

We have to secure a future for our little ones

比与非洲蜜蜂住在一起要好

Better than sharing with the African bees

"与我们这些无刺蜂住在一起，至少比与非洲蜜蜂住在一起要好。他们不像我们这样镇静，当你挡到路时他们会攻击你。我们只是个小蜂群，而且不会待很久。我们甚至能自己建一个独立的入口。"

"At least sharing with us stingless bees is better than sharing with the African bees. They are not as calm as we are and will attack you if you get in their way. We are just a small colony and we won't stay for long. We will even build ourselves a separate entrance."

"我听说过那些有攻击性的非洲杀人蜂。他们将会把我们俩都干掉，也会消灭其他所有与他们相遇的生物。我们最好设法保护自己，以及周围保留下来的生物多样性。也许，你们蜜蜂将会是好邻居……"

"那么，你愿意与我们分享你的筑巢地吗？"

……这仅仅是开始！……

"I have heard about those aggressive African killer bees. They will finish us both off, and everything else that comes in their way. It is better that we protect ourselves, and what's left of the biodiversity around here. Perhaps you bees will make good neighbours after all…."

"So, will you please share your nesting site with us then?"

... AND IT HAS ONLY JUST BEGUN!...

......这仅仅是开始！......

... AND IT HAS ONLY JUST BEGUN! ...

Did You Know?

你知道吗?

There are 20 000 bees species. About 500 species do not sting. Stingless bees will bite when in danger or try to get into the noses and ears of any invaders to deter them.

地球上有 2 万种蜜蜂，其中 500 种不会蛰人。无刺蜂遇险时会叮咬入侵者，或设法钻进它们的鼻子和耳朵中以抵御入侵。

Bee biodiversity is threatened around the world, primarily as a result of the introduction of the more productive African bee that, despite its aggressive behaviour, can produce up to 100 kg of honey per beehive per year.

全世界的蜜蜂多样性正受到威胁，主要是由于引进了更高产的非洲蜜蜂所致。非洲蜜蜂虽然攻击性强，但每个蜂巢每年能产 100 千克蜂蜜。

虽然非洲蜜蜂群体能生产大量的蜂蜜，但单只蜜蜂产量最高的却是无刺蜂。一群约 300 只，生活于沼泽地的麦蜂每年能产出 3 升蜂蜜，这种蜂蜜的价格是普通蜂蜜价格的 10 倍。

While African bees produce large amounts of honey collectively as a colony, the highest individual productivity comes from stingless bees. Melipona marginata living in swarms of about 300 bees, produce 3 litres of honey per year, which is sold at 10 times the price of standard honey.

就像杜鹃这样的巢寄生鸟类在其他鸟巢中产蛋以便让其他鸟类抚养自己的幼鸟一样，也有某些种类的蜜蜂在其他蜜蜂种群（如独居蜂和熊蜂）准备的蜂巢中产卵，因此它们也被称为"杜鹃蜂"。

Just like the parasitic bird, the cuckoo, lays its eggs in another bird's nest for the other birds to raise its chicks, there are bees that lay their eggs in cells prepared by other bee species, such as the solitary or bumble bees, and therefore are called cuckoo bees.

Bees change (metamorphose) like butterflies. A bee passes through four life stages: egg, larva, pupa and adult. The first three stages occur in the brood cell of the nest. Males hatch from unfertilised eggs and females from fertilised eggs.

蜜蜂像蝴蝶一样会发生身体变化，这一过程称为变态。一只蜜蜂需要经历四个生命阶段：卵、幼虫、蛹和成虫，其中前三个阶段发生在蜂巢中的抚幼室。雄蜂从未受精的卵发育而来，而雌蜂从受精卵发育而来。

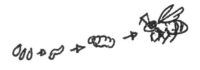

Bees keep most of their stored honey cold so that it is less likely to be eaten by other creatures that do not like cold food. In a beehive heat moves from the outside to the inside, and rises from below to the top, self-warming the honey when the need arises.

蜜蜂把大部分蜂蜜冷藏储存，这样就不会被不喜欢寒冷食物的生物吃掉。有需要时，热量会从蜂巢外部进到内部，从底部上升到顶部，从而使蜂蜜自行升温。

*B*ees collect water and use it to cool the hive through evaporation. They also use water to dilute honey before they consume it. A colony of 50 000 worker bees and 1 000 drones can collect 400 kg of nectar, water, and pollen per year.

蜜蜂采集水，通过水分蒸发给蜂巢降温。它们在吃蜂蜜前会用水稀释。一个由 5 万只工蜂和 1 000 只雄蜂组成的蜂群，每年能采集 400 千克花蜜、水和花粉。

400 千克

*J*ust as surprising perhaps as finding out that there are stingless bees, is learning that there are flightless parrots. The New Zealand night parrot, called a kakapo in the Maori language, is a flightless, nocturnal, ground-dwelling parrot. It is endemic to New Zealand, which means that it occurs only there. It climbs trees and drops down like a parachute.

不能飞行的鹦鹉就像无刺蜂那样令人惊奇。在新西兰毛利语中被称为 kakapo 的鸮鹦鹉，就是一种不能飞行的夜行性地栖鹦鹉，也是新西兰独有的物种。它们能爬树，并像降落伞一样落地。

Think about It

想一想

In your garden, would you prefer bees that can produce honey but do not sting?

在你的花园中，你是不是更喜欢那种能产蜜又不会蛰人的蜜蜂？

如果两种生物正在竞争同一片生存空间，你会建议其中一个退出争斗，还是建议他们设法一起生活呢？

If two creatures were competing for the same living space, would you advise one to give up the battle, or suggest both find a way to live there together?

Who do you blame for the fact that there are now so few trees left in certain regions?

对于在某些地区只剩下如此稀少的树木的现实，你认为该归咎于谁？

拥有良好的睡眠是否比确保后代的未来更重要？

Is having a good sleep more of a priority in life than securing a future for one's offspring?

Is anyone prepared to share his or her home with someone who may be very different: as different as a bee from a parrot? Who is willing to share their bedroom with others who are desperate to find a safe place in which to grow up? Would it make any difference if you were asked to tolerate the presence of others for only a short period, a few weeks for instance? How keen are we to maintain our privacy, and under which conditions are we prepared to make an exception? Make a list of possible exceptions that may motivate people to accept sharing the privacy that seems sacrosanct to them.

　　会有人愿意与自己非常不同的人分享自己的家吗？即便这种差异会大到像蜜蜂与鹦鹉之间那种程度？有人愿意与其他竭尽全力寻找安全成长空间的人分享自己的卧室吗？如果你只被要求忍受很短时间（比如几周），你的选择会有所不同吗？我们是不是很想维护自己的私人空间？会在哪些情况下准备破例呢？某些特殊情况会激励人们作出牺牲去分享私人空间，请列出这些可能情况的清单。

学科知识

Academic Knowledge

生物学	蜜蜂是可以飞的昆虫，与黄蜂和蚂蚁关系密切；麦蜂建立蜂蜡隧道来保护蜂巢入口；花粉和蜂蜜储存在蜂蜡中；蜜蜂只吃植物，充足的食物会刺激蜂王产卵；蜜蜂在1.25亿年前从黄蜂分化而来；有些鹦鹉的雄性和雌性看上去非常相似；鹦鹉寿命长达80—100年。
化 学	无刺蜂的蜂蜜含有高度抗菌的物质；蜂胶是蜜蜂用来密封蜂巢裂缝的一种树脂状物质；工蜂有蜡腺，在它们35天的寿命中，有10—16天在生产蜂蜡；蜡腺将糖转化为片状的蜂蜡，蜂蜡被咀嚼并与唾液混合后变成白色。
物 理	冬季为蜂巢加热时，蜜蜂并不是加热整个蜂巢，最温暖处位于蜂巢中心；为了给蜂巢降温，蜜蜂采集水，拍打翅膀让空气循环；蜜蜂能忍受最高50℃、最低零下45℃的极端温度。
工程学	蜜蜂用自己生产的蜂蜡筑巢，在空间和建材上贯彻了效益最大化原则。
经济学	麦蜂蜂蜜在中美洲和非洲售价很高，当地人把它当作药品，特别是用来治疗眼疾；为了保护当地的无刺蜂，巴西曾经实施禁售，但并未取得成功，现在无刺蜂交易自由，巴西各地的无刺蜂群反而健康生长。
伦理学	巴西法律禁止破坏现有的野生无刺蜂群，但允许转运由蜜蜂自己在野外创建的新蜂群；野生鹦鹉是最常被捕捉和作为宠物交易的动物，导致其野生种群的缩减。
历 史	马德里抄本是仅有的3部保存下来的在哥伦布发现美洲大陆以前的玛雅图书之一，可以追溯到公元900年，描写了古老的玛雅文明对蜂的利用；当亚历山大大帝将鹦鹉从印度引入希腊时，鹦鹉首次在欧洲出现；蜜蜂有1亿年的历史，而鹦鹉有4 000万年历史。
地 理	中美洲的玛雅文明已饲养蜜蜂；马达加斯加拥有很多种无刺蜂。
数 学	蜂群优化算法被用于管理和工程学中；球体表面积与半径的平方成正比，其体积与半径的三次方成正比，这意味着随着蜂群变大，其体积比表面积增加更快，即更大的蜂群以更慢的速度损失热量；蜂窝的开口为六边形，底部四边形的钝角是109° 28′，锐角为70° 32′。
生活方式	蜜蜂越来越多地被当作宠物饲养，尤其是能在都市环境饲养的无刺蜂；野生动物贸易受到《国际濒危物种贸易公约》（CITES）的监管。
社会学	野生鹦鹉从未与猫或老鼠接触，因而未发展出任何反捕食行为，这使它们在引进这些动物的地区有很高的灭绝风险；当大量的幼蜂孵化后，蜂巢就变得过度拥挤，部分蜜蜂将成群飞离蜂巢，寻找新的家园。
心理学	鹦鹉心理学，指鹦鹉容易与人类亲近；鹦鹉的主人更有同情心。
系统论	非本地的非洲蜜蜂不会为所有开花植物授粉，所以引入非本地的蜜蜂物种不仅影响本地蜜蜂物种，还使植物多样性降低。

情感智慧
Emotional Intelligence

鹦鹉

鹦鹉很有礼貌，行为举止像淑女。她平静地指出自己马上就要产蛋了。她的论点前后一致，确信会找到解决方案，不接受任何超过她的能力范围的责任。鹦鹉在争论中非常坚定，表明在仙人掌中筑巢不是她的选择，坚定地回绝了蜂王。鹦鹉重申是她最先到那儿的，因此有权继续待下去。然后她解释道，蜜蜂嗡嗡作响将让共同生活变得几乎不可能。最后，鹦鹉不再解释，开始讨论她们如何共同生存。

蜂王

蜂王的要求很直接，并且在交谈中很坚定，承认鹦鹉先发现那棵树，但强调她的祖先最先出现在那个生态系统中，应该有优先权。蜂王让鹦鹉去找另外一棵树，也意识到对鹦鹉来说这样做极为困难，于是建议鹦鹉把仙人掌作为一种选择。最后，蜂王重申共同生活的建议，因为她知道这在短时间内很有必要。当鹦鹉拒绝蜂王的提议时，蜂王占据道德高地，谈到确保后代未来的重要性。

艺术
The Arts

你的活动选择是，要么用六边形做一个蜂巢模型，要么做一个鹦鹉尾巴的模型。如果你选择做蜂巢模型，就请你算一下需要多少个六边形的"窗户"。如果你选择做鹦鹉尾巴的模型，就用彩色铅笔来画各种各样的羽毛，涂上明亮美丽的颜色。现在，将彩绘的羽毛剪下来，拼接在一起来模拟鹦鹉的尾巴。当然，我们不希望用真正的鹦鹉羽毛来做这个尾巴，因为买卖真正的羽毛将会鼓励非法野生动物产品贸易。

思维拓展
Systems: Making the Connections

森林砍伐通常只与植物的毁灭联系在一起，但我们很少意识到它对许多其他生命形式具有毁灭性的级联效应，威胁这些生物的生存。然而，当森林被单一种植园取代时，野生物种就没有生存空间了。当森林被砍伐殆尽又不种植任何植物时，土地就变成了干旱的热带稀树草原。这会给所有物种带来灾难，因为土地暴露在太阳下，土壤温度升高，偶尔降雨时，也没有植被保护，结果导致大规模的水土流失。冷的雨滴无法穿透坚硬的热土，这种现象称为"煎锅效应"。森林砍伐让大自然很少有机会能再次恢复为有活力的生态系统。少数生存下来的物种，在非常有限的空间里竞争避难所。由于无法保护各种生存要素和摆脱捕食者，即便有能充分满足生存需要的营养，像鹦鹉、蜜蜂这样的物种，生存机会仍然很少。这则童话包含一个特定的哲学反思：资源有限时，我们能在多大程度上接受与他人一起生活带来的不便？特别是与我们并不认识的人，这些人或许与我们截然不同。在环境危机时期，尤其需要社会的凝聚力。它不仅发生在一个家庭或一个物种中，而且发生在所有家庭和同一生态系统的所有物种中。

动手能力
Capacity to Implement

看看我们能做些什么来增加环境中的蜜蜂数量。首先，我们要种植更多的开花植物，找出哪些是本地现有的物种。选择有更多花粉的单花瓣植物，确保你种植的植物有不同颜色的花，因为黄色、白色、蓝色和紫色将比粉色、橙色和红色吸引更多的蜜蜂。选择不同花期的植物组合，这样蜜蜂全年都有食物。同时，选择会为你和家人提供水果和蔬菜的植物。草本植物也会吸引蜜蜂到你的花园来，特别是迷迭香、薰衣草、薄荷、鼠尾草和百里香。

故事灵感来自
This Fable Is Inspired by

德斯蒙德·图图
Desmond Tutu

德斯蒙德·图图大主教的父亲是一位小学教师，母亲在一所盲人学校工作。德斯蒙德喜欢阅读，年轻时最喜欢的书是《伊索寓言》和莎士比亚的戏剧。他于1954年从南非大学毕业，回到自己学习过的中学教英语和历史。他于1961年被任命为牧师，于1966年获得国王学院神学硕士学位。德斯蒙德·图图是首位被任命为约翰内斯堡圣公会教长的黑人，后来短暂地出任主教。他是南非著名的圣公会教士，最著名的是他在反对南非种族隔离制度上的作用。1978年，他被任命为南非教会理事会的秘书长，成为南非黑人权利的主要发言人。

图书在版编目(CIP)数据

冈特生态童书.第三辑修订版:全36册:汉英对照 /
(比)冈特·鲍利著;(哥伦)凯瑟琳娜·巴赫绘;
何家振等译.—上海:上海远东出版社,2022
书名原文:Gunter's Fables
ISBN 978-7-5476-1850-9

Ⅰ.①冈… Ⅱ.①冈…②凯…③何… Ⅲ.①生态环
境–环境保护–儿童读物—汉、英 Ⅳ.①X171.1–49

中国版本图书馆CIP数据核字(2022)第163904号
著作权合同登记号图字09-2022-0637号

策　　划　张　蓉
责任编辑　祁东城
封面设计　魏　来　李　廉

冈 特 生 态 童 书
蜜蜂在家里
[比]冈特·鲍利　著
[哥伦]凯瑟琳娜·巴赫　绘

唐继荣　译

记得要和身边的小朋友分享环保知识哦!
八喜冰淇淋祝你成为环保小使者!